ABOVE: *The twin-cylinder engine at Hesketh Shaft, Chatterley Whitfield Mining Museum, represents the final design phase for deep shaft winders. Drop valve inlet, governor-controlled trip gear, Corliss exhaust valves and bi-cylindro-conical winding drum combine to reduce steam consumption. Built by the Worsley Mesnes Ironworks, Wigan, in 1914.*

COVER: *Colliery winding engine built by Grant, Ritchie of Kilmarnock in 1893, which worked at Lady Victoria Colliery, Newtongrange, Mid Lothian, until 1981. Twin cylinders each 42 inches bore by 84 inches stroke, with Cornish valves and trip gear. Reversing by Gooch link motion.*

STATIONARY STEAM ENGINES

Geoffrey Hayes

Shire Publications Ltd

CONTENTS

Introduction 3
How a steam engine works 5
Pioneers of steam 9
Engine developments 11
Valves and valve gear 21
Boilers 27
Epilogue 29
Engines accessible to the public 30

Copyright © 1979 by Geoffrey Hayes. First published 1979. Second edition 1983, reprinted 1987. Shire Album 42. ISBN 0 85263 652 0.

All rights reserved. No part of this publication may be reproduced or transmitted in any form or by any means, electronic or mechanical, including photocopy, recording, or any information storage and retrieval system, without permission in writing from the publishers, Shire Publications Ltd, Cromwell House, Church Street, Princes Risborough, Aylesbury, Bucks HP17 9AJ, UK.

Printed in Great Britain by C. I. Thomas & Sons (Haverfordwest) Ltd.

ACKNOWLEDGEMENTS
The photograph on page 11 is reproduced by kind permission of the National Railway Museum, York (Crown Copyright). Other photographs and drawings are by the author. The publishers acknowledge with gratitude the assistance and advice of Mr Kenneth Brown in preparing this book for publication.

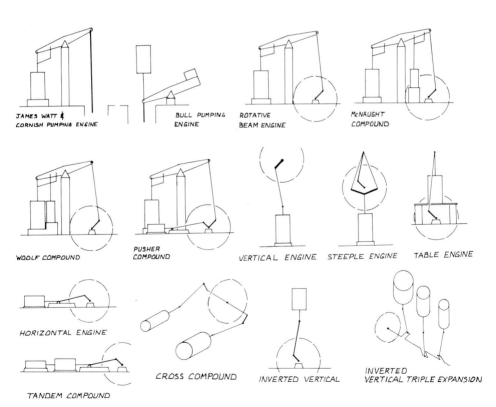

A single-cylinder beam engine pusher compounded by a single-cylinder horizontal engine coupled to the outer end of the crankshaft on the other side of the engine-house wall. The beam engine was built about 1866.

INTRODUCTION

The first practical application of steam power was the beam engine. This form of engine was built over a span of two hundred years. By the nineteenth century the problem of its size and weight was already attracting the attention of inventors. Edward Cartwright, a clergyman, in 1798 devised an engine in which a shaft carrying a flywheel was placed directly above a vertical cylinder and turned by connecting rods without using a beam. Fears of severe wear discouraged builders from using horizontal cylinders and many fascinating designs were produced incorporating vertical cylinders. Eventually these fears were disproved and from about 1830 horizontal engines were built in increasing numbers.

To save space, the inverted vertical engine with cylinders above a crankshaft at floor level was developed towards the end of the nineteenth century. Horizontal and inverted engines became standard types used for every conceivable purpose requiring power. Power was generated on the spot from fuel and water.

Development of the steam turbine and the generation of electricity in central power stations gradually made do-it-yourself power uneconomical. The flexibility and convenience of power arriving down a wire is an invincible force and reciprocating steam power in industry is nearly extinct.

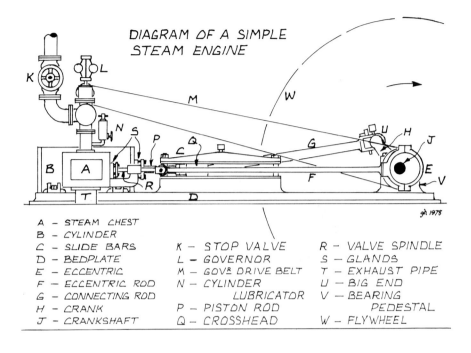

A - STEAM CHEST
B - CYLINDER
C - SLIDE BARS
D - BEDPLATE
E - ECCENTRIC
F - ECCENTRIC ROD
G - CONNECTING ROD
H - CRANK
J - CRANKSHAFT
K - STOP VALVE
L - GOVERNOR
M - GOVR DRIVE BELT
N - CYLINDER LUBRICATOR
P - PISTON ROD
Q - CROSSHEAD
R - VALVE SPINDLE
S - GLANDS
T - EXHAUST PIPE
U - BIG END
V - BEARING PEDESTAL
W - FLYWHEEL

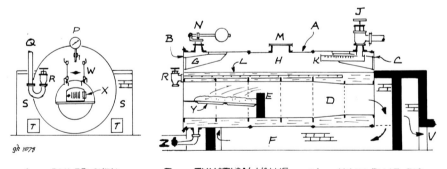

A - BOILER SHELL
B - FRONT PLATE
C - BACK PLATE
D - FURNACE TUBE
E - FIREBRICK BRIDGE
F - BOTTOM FLUE
G - GUSSET STAY
H - STEAM SPACE
K - COLLECTOR PIPE.
J - JUNCTION VALVE
L - WATER LEVEL
M - MANHOLE
N - SAFETY VALVE
P - PRESSURE GAUGE
Q - FEED PIPE
R - CHECK VALVE
S - SIDE FLUES
T - SOOT DOORS
V - MAIN FLUE TO CHIMNEY
W - WATER GAUGES
X - FIRE DOOR
Y - FIRE GRATE
Z - BLOW DOWN VALVE

The beam and parallel motion linkage of a very large Woolf compound beam engine made by Gimson & Co of Leicester in 1884 for sewage pumping at Burton-on-Trent. Massive girders built into the engine-house walls support the beam pivot. The beam, unusually, is a box section made of wrought iron plates.

HOW A STEAM ENGINE WORKS

Take an ordinary garden syringe and look how it is made. There is a tube, inside which slides a disc or *piston*. To the piston is attached a rod which passes out through one end of the syringe, and to the rod is fastened a handle. At the other end of the syringe is a hole or a number of holes. If the syringe is clean, push the handle in and then blow hard down the hole at the end. The handle will then slide out. This is because there is more force or *pressure* on one side of the piston than the other.

Now push in the handle of the syringe and put a thumb over the hole in the end. Pull out the handle — it will be quite hard — and let it go. It will slide in again of its own accord. What has happened is that there is no air on one side of the piston because the thumb has prevented air from entering the syringe. The weight or pressure of air on the handle side of the piston then pushes it along. The effort required to pull out the handle with the hole in the end blocked will give some idea how much pressure the atmosphere can exert.

When the piston of the syringe was pulled along and the thumb over the hole prevented air from entering a *vacuum* was created: no air, no steam, no anything — that is a vacuum.

The ancient Greeks probably knew something about atmospheric pressure and vacuum but it was not really discovered until early in the seventeenth century. About the same time it was also discovered that if water was put in a sealed container or vessel and placed on a fire steam would be generated and with terrific force would blow the vessel to pieces. It was also found that if the vessel full of steam was sprayed with cold water it was likely to collapse with a loud crack. Here was the same effect as the syringe experiment but on a much more powerful scale. So it was discovered that when contained in a vessel steam can exert a great force and when suddenly cooled it turns back into water or *condenses* and a vacuum is formed.

The principle of Newcomen's early steam engines was to allow steam at atmospheric pressure to fill the space underneath a piston in a cylinder while it was lifted by a counterweighted beam. Then the steam was condensed to create a vacuum and the piston was pushed back again by atmospheric pressure acting on top. In those days the air pressure really did the work. Later, Watt admitted steam on top of the piston whilst a vacuum was created underneath. The piston was then moved positively by steam pressure, with assistance from the vacuum.

The early steam engines depended a great deal on the vacuum to work successfully. This was because the vessels or *boilers* in which steam was generated could not be made very strong. As engineering techniques improved and stronger boilers could be made steam pressures gradually increased. Engines then depended less on vacuum and more on steam pressure although vacuum still made a worthwhile contribution to power.

To do useful work the movement of the piston has to be transmitted outside the tube or *cylinder* in which it moves back and forth. Go back to the syringe again. The rod attached to the piston passes out through the end. Exactly the same method is used on a steam engine although the *piston rod* now transmits power from the piston instead of driving it as when the syringe is used to spray water. In the first steam engines the piston rod was attached to one end of a beam, which it rocked up and down. To the other end of the beam was attached either a rod for driving a pump or a *connecting rod* which turned a shaft and had to be heavier than the piston.

More modern engines have the piston rod attached to a *crosshead,* a block of metal guided in a straight line by *slide bars*. To the crosshead is fixed one end of the connecting rod, whose purpose is to convert the sliding or *reciprocating* movement of the crosshead into a turning or rotative movement. To do this the other end of the connecting rod is joined to a *crank*. The crank is fixed to a shaft — the *crankshaft* — which can rotate in *bearings*. Securely fixed on to the crankshaft is the *flywheel,* large and heavy. Its purpose is to even out the pull and thrust of the piston. By fitting a belt or ropes around the rim of the flywheel or by using gearing the engine's power can be transmitted to machinery.

To see how the reciprocating movement of the crosshead is turned into rotative motion, try turning an egg whisk. The handle acts as a crank and one's arm as a connecting rod. The elbow simply moves back and forth.

On an engine the two ends of the connecting rod have different names. The end which fits on the crank is the *big end* and the end which fits on the crosshead is the *little end*.

Some old engines and nearly all *beam engines* have their piston rods guided not by a crosshead and slide bars but by a system of pivoted links called *parallel motion*.

For a steam engine to work steam has to be admitted to the cylinder and allowed to escape or *exhaust* from the cylinder at the right time so that the piston is pushed to and fro. Admission and exhaust are controlled by *valves* operated by a *valve gear* which is driven by the engine itself. There are many forms of valves and valve gears and they will be described in a later chapter.

OPPOSITE: *The single-cylinder beam engine made in Birmingham in 1852 for Combe Sawmill, Oxfordshire, and now preserved. The beam pivot is supported on four inclined columns.*

A Cornish pumping engine made in Cornwall in 1874 (using some older parts) for Prestongrange Colliery, East Lothian, and now preserved as part of the Scottish Mining Museum. Note how the engine-house wall is used to support the engine beam and that there is no crank : the engine worked the pump rod directly.

The beam of a Cornish-type pumping engine and the cylinder top of a Bull pumping engine at Kew Living Steam Museum.

PIONEERS OF STEAM

Thomas Savery, an inventor who lived in Devon, knew about pressure and vacuum. In 1698 he succeeded in making a device which pumped water by steam. His invention was not strictly an engine as it had no moving parts. Steam was blown into a *receiver* and then condensed by water being sprayed over the receiver to create a vacuum. The water to be pumped was then drawn into the receiver. When it was full, steam was turned on and blew the water out of the receiver up a pipe to a higher level. Water is heavy and to do useful work the steam had to be at high pressure. The engines often exploded.

Thomas Newcomen, an ironmonger at Dartmouth, often visited the tin and copper mines in Cornwall in the course of his business. To drain their mines the miners used pumps worked by men or horses. Newcomen visualised the piston sliding in a cylinder. He condensed the steam by spraying water into the cylinder. Only low steam pressure was needed for the engine to work.

Newcomen put his engine at the side of the mine shaft and connected it to the pumps by chains attached to an overhead rocking beam. The first engine was put to work at a coal mine near Dudley Castle (West Midlands) in 1712. An actual Newcomen engine and a model of one can be seen in the Science Museum, London.

Newcomen engines used a lot of coal because the cylinder was cooled by water and had to be heated up again at each stroke of the piston. James Watt, who worked at Glasgow University, realised this when he was given a model of a Newcomen engine which would not work. After many experiments he found that if after doing its work in the cylinder the steam was led to a separate chamber and condensed there the cylinder would remain hot all the time. As the cylinder was connected to the *separate condenser* by a pipe the steam rushed from the cylinder to the condenser and a vacuum was created in the cylinder.

By about 1776 James Watt had a pumping engine working successfully. It used much less coal than a Newcomen engine. Watt admitted steam to one side of the piston while maintaining a vacuum on the other. To make sure there was always a vacuum in the condenser he fitted a small pump driven by the engine to remove air which leaked in. This *air pump* also removed the water formed by the condensing steam. Before 1780 all steam engines were non-rotative, that is they imparted a reciprocating motion directly to pumps in a well or shaft. But in 1783 Watt introduced the rotative *double acting engine* in which steam was admitted and exhausted on each side of the piston in turn, pushing it to and fro with positive action.

James Watt engines may be seen in the Science Museum and the Royal Scottish Museum at Edinburgh.

A beam engine is large and heavy. Edward Bull, a foreman engine erector for the firm of Boulton and Watt in Cornwall at the end of the eighteenth century, devised a mine pumping engine without a beam. The cylinder was inverted directly over the mine shaft and the piston rod coupled straight on to the pump rods. *Counterweights* on a balancing beam were provided if the pump rods were heavy. Cheaper than a beam pumping engine, the Bull engine was the first major departure from the beam engine design. Bull was prosecuted by James Watt for infringement of his patent and only a few engines were erected in Cornwall. Later, the Bull engine was quite widely used for pumping at collieries. A Bull engine can be seen at Kew Living Steam Museum.

Edward Cartwright's vertical engine has already been mentioned. There is a model of this engine in the Science Museum but Cartwright's engine never came into use. In 1800 Phineas Crowther took out a patent for a *vertical engine*, the design of which gained wide acceptance. It was used for colliery winding engines with great success. Engines of the vertical type were in use until quite recently.

At the North of England Open Air Museum at Beamish, Co. Durham, there is a colliery winding engine of the Crowther type and Brighton and Hove Engineerium has a small vertical mill engine.

A crankshaft with a heavy *flywheel* rotating at speed causes vibration if not well supported and the large vertical engines needed a heavy structure of stone, cast iron or timber if they were to run smoothly. *Table engines* and *steeple engines,* invented in the early nineteenth century, represented attempts to lower the height of the crankshaft. Examples of these engines can be seen in the Science Museum and the Royal Scottish Museum.

George and Robert Stephenson, when building the Leicester and Swannington Railway in 1833, had to carry their line down a hill to a colliery at Swannington and they put a winding engine at the top of the hill to wind up the wagons by a rope. Whereas nearly all inventors and engineers made their engines with vertical cylinders this engine had a *horizontal* cylinder. The Swannington engine worked from 1833 to 1946 and is now retired in the National Railway Museum, York.

This development marked the end of the pioneer era. From that time was a period of continuous improvement. There were many inventions still to come but these were concerned with improving the reciprocating engine while retaining the form which had become established.

Swannington Incline winding engine. Note the 'tail rod' passing out of the rear of the cylinder and guided by slides. The tail rod helped to support the piston. Made in 1833 and the oldest known horizontal engine surviving, it is now in the National Railway Museum, York. (Crown Copyright)

ENGINE DEVELOPMENTS.

The Stephensons and other locomotive engineers with their use of horizontal cylinders showed that excessive wear along the bottom of the cylinder did not take place.

Large beam engines and vertical engines were supported by the walls of their engine houses, so the walls had to be very thick and the engine houses were built like castles. The horizontal engine needed only a foundation or *bed* of strong stones. The engine house was completely separate and the walls could be made much thinner. Cost of building was much reduced.

For a time beam, vertical and horizontal engines developed side by side, but table and steeple engines dropped out completely. Improvements in engineering meant that smaller rotative beam and vertical engines could be made self-contained. Cast-iron frames of A shape or a number of Grecian columns supported the beam pivot or, in the vertical engine, the crankshaft. An A-frame beam engine can be seen at Pinchbeck Marsh near Spalding in Lincolnshire.

Rotative beam engines were popular for pumping in the water-supply and sewerage schemes which were constructed in cities and towns in the nineteenth century. Many of these engines worked until recently. Beam engines pumped sewage at Burton-on-Trent until 1971. Engines may be seen at Papplewick near Nottingham, Abbey Pumping Station at Leicester, Hove Engineerium, Kew and Ryhope Pumping Station near Sunderland.

The *Cornish* non-rotative pumping engine evolved for keeping the deep Cornish tin mines clear of water; it became renowned for its efficiency and was used all over the world. Superb

LEFT: *Twin-cylinder vertical A frame engine with slide bars guiding the crosshead and disc cranks, geared to a centrifugal pump for sedge moor drainage; Meyer expansion slide valves, non-condensing exhaust to atmosphere; Eastons Amos & Anderson, London, 1869. Preserved in situ, Aller Moor pumping station, Burrow Bridge, Somerset.* RIGHT: *Cylinders and parallel motion linkage of a small Woolf compound beam engine at a Kent waterworks. The engine is self-contained with the beam pivot supported on a single Grecian column.*

examples of Cornish engines can be seen at Kew, Crofton, East Pool, Hull and Prestongrange Mining Museum near Edinburgh.

Arthur Woolf, early in the nineteenth century, knew that when steam was exhausted from one cylinder it still had power left in it. He took the exhaust steam into a second and larger cylinder where it was able to push the piston with reduced but still useful force. The first cylinder he called the *high-pressure* cylinder and the second and larger cylinder the *low-pressure* cylinder. At the time of this invention steam pressures were still very low and it was not very successful, but in time, with steam at higher pressures, the system was widely adopted. Engines built on this system are called *compounds*. The beam engines at Leicester, Hove and Ryhope are built on the *Woolf compound* principle.

From 1850 onwards the horizontal engine came into favour more rapidly for driving mills and winding coal at deep colliery shafts. At first with low steam pressures horizontal engines were built with single cylinders. If more power was needed two cylinders were provided. Each cylinder drove a common crankshaft with the flywheel mounted in the middle. Twin engines were very popular for colliery winding as they started very easily, and the winding drum took the place of the flywheel.

When the compound principle was adopted on horizontal engines it took two main forms: the *cross compound* or twin engine principle, where the cylinder on one side exhausted to the cylinder on the other, and the *tandem compound*. Here the high and low pressure cylinders were

LEFT: *Woolf compound A frame beam engine made in 1835 by Wentworth & Sons, Wandsworth, still at work in 1978 driving the machinery at a London brewery.* RIGHT: *Cross-compound vertical 'house-built' engine with parallel motion guiding crossheads; governor-controlled expansion slide valve on high-pressure cylinder, simple slide valve on low-pressure cylinder; exhaust to condenser below floor. Built about 1860, it drove a Yorkshire woollen mill through two flat belts from the flywheel.*

placed one behind the other. A long piston rod passed through the two cylinders to connect the pistons together. An old tandem compound built about 1873 may be seen at Glenruthven textile mill at Auchterarder in Perthshire and a more modern one, made in 1900, at the Gladstone Pottery Museum, Longton, Stoke-on-Trent.

Where very high powers were required twin tandem compound engines were often used, consisting of two complete tandem engines coupled together by a common crankshaft. The flywheel was mounted on the crankshaft centrally between the engines. Two of these engines may be seen at the Tower Bridge Museum, London E1.

Ships always had the problem of requiring high power with little space to put the engine. Marine engineers developed the *inverted vertical engine* in which the crankshaft was at floor level and the cylinders were supported on columns above. During the last quarter of the nineteenth century this type of engine rapidly came into favour for land use. It was built in an enormous range of sizes. Usually these engines ran quite fast but because they occupied so little space they were also popular as slow-speed waterworks pumps.

As steam pressures were becoming higher the steam was often passed through three cylinders before exhausting. These engines were called *triple expansion.* One is on view at Broomy Hill Waterworks, Hereford, and a compound engine at Diamond Works, Royton, near Oldham. Because of their ancestry these engines are often called *marine type* but none of them had reversing gear or ever drove a ship.

Single-cylinder horizontal engine with drop valves and governor-controlled trip gear, which drove a generator at a Northamptonshire gelatine works. It was made in France by Messrs Piguet of Lyon-Anzin. Non-condensing exhaust to low-pressure process steam mains.

Around 1900 electricity was coming into use and the machines — dynamos or generators — which made the electricity ran at high speed. They were often coupled directly to the end of the crankshaft of an inverted vertical engine, running at high speed, with the moving parts enclosed in an oil bath. Many of these enclosed engines were made and are still made today. Engines of this type are displayed at the Prestongrange Mining Museum and the Birmingham Museum of Science and Industry.

The final reciprocating engine development was the *Uniflow* type, perfected by Professor Stumpf. Steam was admitted at each end of the cylinder in turn but was exhausted from the centre of the cylinder. The piston itself uncovered *ports* in the cylinder through which the exhaust steam escaped. This form of engine was usually single-cylinder horizontal with all the parts enclosed. It ran at speeds of up to 150 rpm and was very efficient. A Uniflow engine made by W. & J. Galloway of Manchester is in the Museum of Science and Industry, Birmingham.

The types of engines so far described formed the majority of those built, but some engines did not condense their exhaust steam. They were so made either for cheapness or where starting and stopping were frequent, as with colliery winding and rolling mill engines. Some factories needed much steam at low pressure for their processes and exhaust steam from the engine was used.

A number of special types were made for pumping and very numerous were the *direct acting* pumping engines, in which the piston rods were coupled to the pump directly or via bell cranks and there was no crankshaft. These engines had ingenious valve gears worked from the piston rods.

There were also many small rotative pumps with the piston rod shaped like a banjo or guitar. The crank and connecting rod rotated *inside* the banjo. These were known as *Cameron* or *Mumford* pumps after the firms which made most of them.

Many old steam engines were very strongly built. When more power was needed from a large single-cylinder engine it was often possible to fit a second, smaller cylinder which took in steam at high pressure and exhausted into the old cylinder. This converted the engine into a compound and power output was often doubled.

James McNaught of Glasgow converted many single beam engines to his own system of compound. Others were converted by adding a horizontal high-pressure pusher cylinder and crank. Many single-cylinder horizontal engines were converted into tandem compounds.

ABOVE: *Horizontal tandem compound engine with simple slide valves and throttle governor. Exhaust to condenser seen alongside cylinders; boiler feed pump in foreground. Air pump and feed pump driven by side rod from crosshead. Built about 1873, it drove Glenruthven textile mill until 1980. Note the flat driving belt. Operated at 64 rpm.*
BELOW: *Single-cylinder horizontal engine with expansion slide valves and governor-controlled link motion. Used to drive a Berkshire sawmill. Built by Ransomes Sims & Jefferies, 1899. Note 'trunk' crosshead guides and disc crank. Non-condensing exhaust to atmosphere.*

ABOVE: *Cross compound horizontal engine with Corliss valves on high-pressure cylinder and simple slide valve on low-pressure. Built 1910 by Clayton Goodfellow of Blackburn, it drove a Lancashire cotton mill until 1973. The rope drive to lineshafting was typical. Each rope would transmit about 40 horsepower (30 kW). The slots in the flywheel are to take a pinch bar for 'barring' the engine to starting position.*
BELOW: *Tandem compound horizontal engine with drop valves built by James Carmichael in 1923; it drove a Scottish sawmill until 1975. Drive to mill by cotton ropes from flywheel at 80 rpm.*

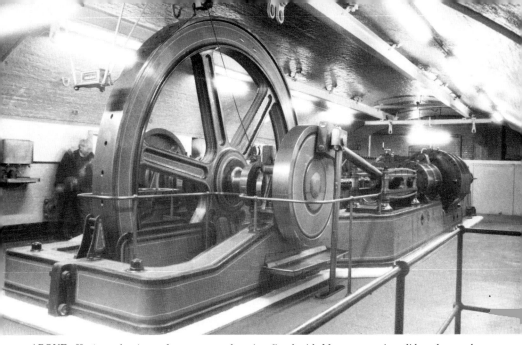

ABOVE: *Horizontal twin tandem compound engine fitted with Meyer expansion slide valves and driving force pumps by the tail rods. Water was supplied at 750 psi to hydraulic machinery working the bascules of Tower Bridge. Built by Armstrong Mitchell of Gateshead, 1894.*

BELOW: *Pollit & Wigzell patent three-piston rod tandem compound engine. Built 1886, fitted with new Corliss valve high-pressure cylinder in 1931. Simple slide valve on low-pressure cylinder. Exhaust to condenser behind low-pressure cylinder. Tail rod drive to air pump. Standby power at a Yorkshire woollen mill in 1976. Gear drive to mill concealed behind panelling. It operates at 56 rpm.*

ABOVE: *A small tandem compound engine built by Marshall & Sons in 1907 and fitted with drop valves and trip gear. Installed in a fen drainage station in Lincolnshire, the engine was direct-coupled to a centrifugal pump, delivering 20,000 gallons (90,000 litres) of water per minute at 140 rpm.*
BELOW: *Cross compound horizontal engine with Meyer expansion slide valves on both cylinders; built by Worthington-Simpson in 1915 for a waterworks near London, two engines drive two well pumps from the low-pressure tail rod and a force pump from the high-pressure tail rod.*

Inverted vertical compound engine with Corliss valves on high-pressure cylinder and piston valve on low-pressure. Exhaust to condenser, the air pump being driven by a rocking beam from the low-pressure crosshead. Built by Scott & Hodgson, 1912, for Diamond Rope Works, Royton, and now preserved in situ. Note the rope 'race' taking drive to several floors.

ABOVE LEFT: *Tiny inverted vertical single-cylinder high-speed enclosed engine direct-coupled to a 2.5 kilowatt generator. Ran at 500 rpm and provided electricity for lighting in a Norfolk waterworks. Made by Dodman of King's Lynn.* ABOVE RIGHT: *Small inverted vertical single-cylinder engine by Marshall & Sons. This was a standard catalogue design available in several sizes.*

BELOW: *A very large inverted vertical triple expansion pumping engine with drop valves built by Worthington-Simpson for Kempton Park works of the Metropolitan Water Board in 1928. Delivering 20 million gallons (90 million litres) per day, the engine developed 1000 horsepower (745 kW) at 24 rpm. Standing 62 feet (19m) high from pump base to top of cylinders it and its twin in the same house were believed to be the largest working steam engines in the world when superseded in 1980.*

Dobson's Corliss valve gear on the triple expansion inverted vertical pumping engine, built by Ashton Frost in 1911, at a Nottinghamshire waterworks. Corliss valves are fitted on all cylinders; this is the low-pressure cylinder with hand-adjusted trips. The engine drove two borehole pumps at 160 feet (49 m) below surface and three force pumps below the engine bed. It delivered 3.5 million gallons (15.9 million litres) a day.

VALVES AND VALVE GEAR

For a steam engine to work, steam has to be admitted and exhausted from the cylinder at the right time. Most engines are *double acting*, that is steam acts on each side of the piston alternately. Admission and exhaust are controlled by *valves* moved by a *valve gear*, which is operated by the engine itself. Early valves were shaped like a bath plug and dropped into a matching hole or *seat* to close. Steam pressure helped to keep the valves tightly closed but also made them hard to open. To see the effect, when next emptying a bath lift the plug only slightly. Water will flow past but when the chain is let go the plug will drop into the hole.

William Murdoch, chief engine erector for the firm of Boulton and Watt, invented the *slide valve*. This consists of a shoe inside a box or *steam chest*. When moved to and fro by the *valve gear* the slide valve uncovers slots or *ports* which communicate with each end of the cylinder. Steam is thus admitted from the steam chest to each end of the cylinder alternately and allowed to escape to exhaust. Much less effort was needed to move a slide valve than to lift valves of the bath-plug type against steam pressure.

Cornish engineers, with their very large mine pumping engines and no rotating shaft to work an eccentric, tackled the problem in a different way. They took two 'bath plugs' and fixed them together a little distance apart. The two valves dropped into two seats to close. In the closed position steam pressure pushed down on the top valve but pushed up on the bottom valve. It was therefore balanced and needed little effort to open. This type of valve was often called the *Cornish valve* or the *double-beat drop valve*. Engines with these valves can be seen at Kew, Papplewick and Ryhope.

Henry Corliss, an American, invented a *semi-rotary* or *rocking valve* in 1849. This valve was very similar in shape to a rolling pin, with a large piece gouged out of the middle. The valve was fitted into a sleeve. Rolling a piece of paper into a tube and slipping it over a rolling pin will give some idea of how the valve was fitted. Ports in the sleeve communicated with the cylinder. A little crank on the end of the valve gave it a twisting movement when waggled by the valve gear. This movement covered and uncovered the ports and admitted and *cut off* the steam. A separate valve did likewise with the exhaust steam. Four valves were needed, two at each end of the cylinder, one for admitting steam and one for exhaust. The Corliss valve was gradually taken up by British engine builders and became widely used.

A development of the slide valve was the *piston valve*. This consisted of two metal discs fixed some distance apart on a spindle. The assembly was made a close fit in a cylindrical steam chest. Steam pressure pushed at each end of the valve and it was in balance. Very little effort was needed to move it to and fro. The earliest known piston valve is on the Swannington incline engine in the National Railway Museum, but piston valves did not come into general use until near the end of the nineteenth century. The piston valve worked in exactly the same way as a slide valve. It was much used on high-speed engines, locomotives and winding engines.

It is likely that the early experimental engines of Newcomen had hand-operated valves. The engine at Dudley Castle, however, had a valve gear. The valves had long handles to open and close them and these engaged with pegs on a wooden rod or *plug rod* hung from the engine beam. Once the engine had been started the plug rod rose and fell with the beam and the pegs struck the valve handles in sequence and the engine kept on working. This *plug-handle* valve gear was used on all non-rotative beam engines. Go to Crofton or Kew Bridge to see it in action.

Plug-handle valve gear was used on early rotative engines but it was soon found more convenient to work the valve gear from an *eccentric*. An eccentric is an iron disc fixed to the crankshaft or on a shaft driven by the crankshaft. The shaft does not go through the centre of the disc but to one side. As the shaft rotates, the disc wobbles round as the wheels on a circus clown's bicycle. This wobbling motion can be transmitted by rods as a to and fro motion working a valve or valves.

To control the speed of rotative steam engines Watt fitted a *centrifugal governor*. Its principle is easily demonstrated. Tie a small but heavy object to a length of string. Make sure there is plenty of space around and holding one end of the string begin to whirl the object round. As the whirling becomes faster the object will rise.

Watt used two heavy spheres suspended on metal arms. The governor was rotated by the engine, which, as it went faster, whirled the spheres round faster. As the spheres rose, the arms to which they were attached moved a link, which in turn was connected to a valve in the steam pipe to the engine. The valve was gradually closed, *throttling* the steam supply. If the engine went slower the spheres dropped and the valve opened wider. This form of governor was widely used on rotative engines.

When the throttle valve was closed by the governor steam had to fight its way past the valve and lost pressure. When it reached the cylinder it gave the piston a gentle push. As steam pressures became higher it was found that if the steam gave the piston a sharp kick and was then cut off smartly it would continue to push the piston but using much less steam in the process.

Various valves and valve gears were devised so that while steam was being admitted to the cylinder all valves were wide open. Very common was Meyer's *expansion slide valve*. This comprised two slide valves, each driven by its own eccentric and riding one on the back of the other. The upper slide valve controlled the instant of cut-off. The simplest expansion slide valves had a wheel to adjust the cut-off. More sophisticated types were driven by a *link motion* which the governor adjusted so that steam was cut off earlier or later.

Corliss and drop valves could be driven by governor-adjusted *trip gear*. The *trips* released the steam inlet valve, which was then closed quickly by a spring.

RIGHT: *Types of cylinders*

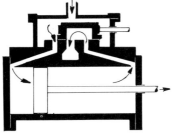

SLIDE VALVE CYLINDER

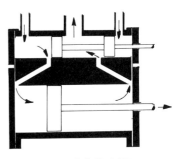

PISTON VALVE CYLINDER

BELOW: *Drop valve gear on a Hathorn Davey triple expansion waterworks pumping engine. The valves are lifted by a special form of cam with a die engaging the valve stem. This recedes quickly to minimise wear.*

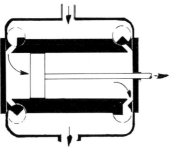

CORLISS VALVE CYLINDER

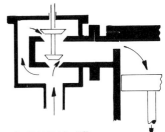

CORNISH OR DOUBLE BEAT DROP VALVE

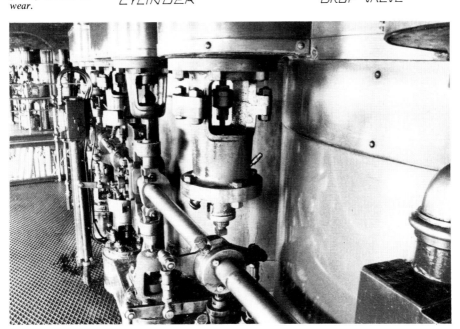

ABOVE: *Triple expansion inverted vertical engine fitted with Morley's patent piston drop valves; built by Thames Ironworks in 1911, for a Metropolitan Water Board pumping station in Surrey; a quick running engine direct-coupled to a centrifugal pump delivering 30 million gallons (136 million litres) per day at 135 rpm.*
BELOW: *Compound enclosed high-speed engine fitted with piston valves and direct-coupled to a 25 kilowatt direct current generator. It ran at 450 rpm and supplied lighting at a London waterworks. Built by Ashworth & Parker of Bury, 1924.*

ABOVE: *Triple expansion duplex pumping engine built by Worthington-Simpson in 1938 for a Norfolk waterworks; used until 1976 supplying 3 million gallons (14 million litres) a day to Wisbech. It ran at 33 strokes per minute and was fitted with Corliss valves on high-pressure cylinders and semi-rotary slide valves on the others. The engine is direct acting and has no crankshaft or flywheel.*
BELOW: *Enclosed high-speed piston valve compound engine coupled to a 14.5 kilowatt alternator running at 600 rpm. It supplied lighting to a waterworks on Romney Marsh; made by Reader of Nottingham.*

ABOVE: *At Kempton Park the 1928 engine-house (left) was built in accordance with contemporary architectural ideas. The chimney built alongside an earlier chimney was carefully designed to match the Edwardian style of the earlier engine-house (right).*
BELOW: *Horizontal Uniflow engine direct-coupled to a centrifugal pump; one of four built in 1923 by Worthington-Simpson and installed in a London waterworks. Each unit could raise 75 million gallons (340 million litres) per day from the river Thames to a raw water reservoir at 133 rpm.*

Range of three Cornish boilers supplying steam to Worthington-Simpson pumping engines. The boilers were fitted with superheaters and economisers. Note the boiler fittings. The black and white stripes behind the water gauges are an aid to seeing the water level.

BOILERS

All steam engines need one or more boilers to provide steam to make them work. The first boilers were merely closed tanks of water set in brickwork formed into *flues* which led the flames and smoke from the fire underneath and around the sides of the boiler before passing to the chimney. The boilers which supplied steam to the Newcomen and Watt engines were usually *haystack* or *wagon* type, their names describing their shape. The wagon boiler resembled the wagons seen in western films. These boilers were unsafe if worked at more than a few pounds per square inch steam pressure. Later came the *egg-ended* boiler, shaped like a long thin straight sausage. It was rather stronger. The firegrate was underneath at the front. Some can now be seen in use as water tanks.

The need for higher steam pressures and fuel economy in pumping at the Cornish mines led to the invention of the *Cornish boiler* by Richard Trevithick. It was first used about 1812. Similar in proportion to a large tin of baked beans lying on its side, it had a tube running through its length. The tube was about half the diameter of the outer shell and in one end was fixed the firegrate or furnace. As well as containing the fire and directing the flames and smoke to the back of the boiler the furnace tube helped to support the ends of the boiler. Brick flues directed the flames and smoke issuing from the furnace tube down and under the boiler to the front. Here they were split and passed along each side of the boiler to the rear again before escaping to the chimney.

In 1844 William Fairbairn of Manchester brought out the *Lancashire boiler*, which had two furnace tubes, each about one third the diameter of the outer shell. Cornish and Lancashire boilers with some variations were the principal boilers used to drive steam engines.

Cornish and Lancashire boilers can be seen at the various sites where engines are

Two Lancashire boilers 7 feet by 28 feet (2.1 by 8.5 m), which supplied steam at 200 psi to two triple expansion duplex pumps. Fitted with forced draught furnaces, mechanical stokers, superheaters and Green's economisers. Boilers by Dodman of King's Lynn, 1938; stokers by James Proctor of Burnley.

preserved. Both types of boilers are very strong and long lasting. Some boilers in use today are over eighty years old. A typical Lancashire boiler would be 8 feet (2.4 metres) diameter by 30 feet (9.1 metres) long with a working steam pressure of 120 pounds per square inch (8437 grams per square centimetre).

Boilers have a number of fittings vital to their working. On the front are the *water gauges*, which show by means of glass tubes the level of water inside the boiler, the *pressure gauge*, which indicates the steam pressure, and the *check valve*, which controls the flow of *feed water* into the boiler. Water has to be fed into the boiler as fast as the water in the boiler is evaporated into steam. On top of the boiler is the *safety valve*, which blows off steam should the pressure become too high, the *junction valve*, which controls the flow of steam from the boiler, and the *manhole*, which allows access into the boiler for inspection and cleaning.

Various devices were used to reduce fuel consumption. One of these was the *economiser*, invented by E. Green. This comprised a number of cast iron pipes set vertically in the main flue to the chimney. Feed water on its way to the boiler was pumped through these pipes and picked up heat from the hot gases leaving the boiler.

Another device was the *superheater*. Here steam on its way from the boiler to the engine was passed through a battery of small tubes fitted in the boiler flue. The heat in the smoke and flames thoroughly dried out the steam and made it hotter. Engines work best if the steam is hot and dry, providing the cylinders are properly lubricated.

To reduce labour and increase efficiency some boilers were fitted with *mechanical coal stokers*. These machines fed coal on to the fires at a rate dependent on the steam demand. More recently boilers have been converted to burn oil or gas. See the gas-fired Lancashire boilers at Kew as an example.

Breweries and whisky distilleries used a large number of small engines. This single-cylinder horizontal engine drove machinery at the Glenmoray-Glenlivet distillery. Fitted with a throttle governor and simple slide valve, it was built by G. Chrystal, St John's Foundry, Perth, about 1897.

EPILOGUE

In these days of universal electric power, not only in industry but in the home, it is difficult to realise that in Victorian times almost the only source of mechanical power was the reciprocating steam engine. It was used much as the electric motor is used today. Thousands were made and in comparison very few survive and even fewer are still at work. These engines did their work behind closed doors, unseen except by those who worked with them or those few who were interested enough to obtain permission to visit the engine house. Today, although so few engines remain, some can now be seen. A number of engines are displayed in museums but, better still, following the efforts of a few individuals and with the co-operation of some public and private bodies, engines can be seen in occasional operation.

The steam engine, however, is not completely dead. The steam turbine still provides the great bulk of the world's demand for electricity. Even the time-honoured reciprocating engine can still be economic in low power ranges.

In food and other industries where steam is required at low pressure for processes, power requirements are moderate and waste material is available for fuel, the steam engine can now show definite advantages over other forms of power. Belliss & Morcom of Birmingham have recently supplied more than fifty of their compound high-speed enclosed engines to various parts of the world. It is fitting that steam engines are still being made in the country of their birth.

ENGINES ACCESSIBLE TO THE PUBLIC

Many of the locations listed are privately owned and operated. The hours of opening may be restricted to weekends or even to certain days in the year. An asterisk at the end of an entry denotes that an engine can sometimes be seen running under steam. Two asterisks denote engines run weekly or daily. Times of opening should be verified before visiting.

ENGINES IN MUSEUMS
Bass Brewing Museum, Horninglow Street, Burton upon Trent, Staffordshire. Telephone: Burton upon Trent (0283) 45301*.
Biggar Gas Works Museum (Royal Museum of Scotland), Biggar, Lanarkshire. Telephone: 031-225 7534.*
Black Country Museum, Tipton Road, Dudley, West Midlands. Telephone: 021-557 9643.*
Blackgang Chine Theme Park, Ventnor, Isle of Wight PO38 2HN. Telephone: Ventnor (0983) 730 330.**
Blists Hill Open Air Museum, Ironbridge Gorge Museum Trust, Telford, Shropshire. Telephone: Ironbridge (095 245) 3522.**
Bolton, Central Precinct, Bradshawgate, Bolton, Lancashire (in glass engine house).
Bolton Mill Engine Museum (Northern Mill Engines Society), Atlas Number 3 Mill, Chorley Old Road, Bolton, Lancashire.*
Bolton Museum and Art Gallery, Le Mans Crescent, Bolton, Lancashire BL1 1SA. Telephone: Bolton (0204) 22311 extension 379.
Bradford Industrial Museum, Moorside Road, Eccleshill, Bradford. Telephone: Bradford (0274) 631756.
Bressingham Gardens and Live Steam Museum, Diss, Norfolk. Telephone: Bressingham (037 988) 386.
Bridewell Museum of Local Industries, Bridewell Alley, Norwich. Telephone: Norwich (0603) 22233.
Bristol Industrial Museum, Princes Wharf, Bristol. Telephone: Bristol (0272) 299771.
Bygones at Holkham, Holkham-next-the-Sea, Norfolk. Telephone: Fakenham (0328) 710806.
Calderdale Industrial Museum, Square Road, Halifax. Telephone: Halifax (0422) 59031.*
Cambridge Museum of Technology, Riverside, Cambridge. Telephone: Cambridge (0223) 68650.*
Cheddleton Flint Mill, Cheddleton, near Leek, Staffordshire.
Cothey Bottom Heritage Centre, Westridge Leisure Complex, Ryde, Isle of Wight.
Darwen, Lancashire; outside India Mill, Bolton Road, and public gardens, Blackburn Road.
Forncett Industrial Steam Museum, Forncett St Mary, Norfolk.*
Gladstone Pottery Museum, Uttoxeter Road, Longton, Stoke-on-Trent. Telephone: Stoke-on-Trent (0782) 311378.
Golden Lion Home Brewery, High Street, Southwick, Hampshire.*
Greater Manchester Museum of Science and Industry, Liverpool Road, Castlefield, Manchester M3 4JP. Telephone: 061-832 2244.**
Hamilton Museum, 129 Muir Street, Hamilton, Lanarkshire. Telephone: Hamilton (0698) 283981.
Higher Mill Museum, Holcombe Road, Helmshore, Lancashire. Telephone: Rossendale (070 62) 226459.
Hollycombe House, Liphook, Hampshire. Telephone: Liphook (0428) 723233.*
Holroyd Gear Works (in glass engine house), Rochdale, Lancashire.
Kirkcaldy Industrial Museum, Forth House, Abbotshall Road, Kirkcaldy, Fife. Telephone: Kirkcaldy (0592) 60732.
Launceston Steam Railway, Old Gasworks, Newport Industrial Estate, Launceston, Cornwall. Telephone: Launceston (0566) 5665.*
Leeds Industrial Museum, Armley Mill, Leeds. Telephone: Leeds (0532) 637862.
Monks Hall Museum, 42 Wellington Road, Eccles, Lancashire. Telephone: 061-789 4372.
Museum of East Anglian Life, Abbots Hall, Stowmarket, Suffolk. Telephone: Stowmarket (044 92) 2229.
Museum of Lincolnshire Life, Barton Road, Lincoln. Telephone: Lincoln (0522) 28448.

Museum of Science and Engineering, West Blandford Street, and Exhibition Park, Newcastle upon Tyne. Telephone: Newcastle (0632) 815129.*
Museum of Science and Industry, Newhall Street, Birmingham. Telephone: 021-236 1022.*
National Mining Museum, Lound Hall, Haughton, Retford, Nottinghamshire. Telephone: Mansfield (0623) 860728.*
National Portrait Gallery (Gallery 14), St Martin's Place, London. Telephone: 01-930 1552.
National Railway Museum, Leeman Road, York. Telephone: York (0904) 21261.
Newcomen Engine House, Royal Avenue Gardens, Dartmouth, Devon. Telephone: Dartmouth (080 43) 2923.
North of England Open Air Museum, Beamish, near Stanley, County Durham. Telephone: Stanley (0207) 31811.*
Nottingham Industrial Museum, Courtyard Buildings, Wollaton Park, Nottingham. Telephone: Nottingham (0602) 284602.*
Penrith Steam Museum, Castle Street, Penrith, Cumbria.**
Poldark Mining Ltd, Wendron, near Helston, Cornwall. Telephone: Helston (032 65) 3531.
Royal Museums of Scotland, Chambers Street, Edinburgh. Telephone: 031-225 7534.
Salford Museum of Mining, Buile Hill Park, Eccles Old Road, Salford. Telephone: 061-736 1832.
Science Museum, Exhibition Road, South Kensington, London. Telephone: 01-589 3456.*
Sheffield Industrial Museum, Kelham Island, Alma Street, Sheffield. Telephone: Sheffield (0742) 22106.**
Somerset County Museum, Taunton Castle, Taunton, Somerset. Telephone: Taunton (0823) 3451.
Staffordshire County Museum, Shugborough, near Stafford. Telephone: Little Haywood (0889) 881388.
Strumpshaw Hall Museum, Strumpshaw, near Norwich. Telephone: Norwich (0603) 712339.
Thursford Collection, Thursford, Fakenham, Norfolk. Telephone: Fakenham (0328) 3839.*
Tolgus Tin Mining Museum, Redruth, Cornwall. Telephone: Redruth (0209) 215171.
Tolson Memorial Museum, Ravensknowle Park, Huddersfield. Telephone: Huddersfield (0484) 41455.
Ulster Museum, Botanic Gardens, Belfast. Telephone: Belfast (0232) 668251.
Welsh Industrial and Maritime Museum, Bute Street, Cardiff. Telephone: Cardiff (0222) 371805.*
Wortley Top Forge, Thurgoland, near Barnsley, South Yorkshire.*

ENGINES PRESERVED IN SITU
Abbey Pumping Station (Leicestershire Museum of Technology), Corporation Road, Leicester. Telephone: Leicester (0533) 61330.*
Aller Moor Pumping Station (Wessex Water Authority), Burrowbridge, Somerset. Telephone: Burrowbridge (082 369) 324.
Astley Green Colliery (Red Rose Live Steam Group), near Leigh, Lancashire. Telephone: Leigh (0942) 878981.
Bamford Mill (Carbolite Furnaces Ltd), Bamford, near Castleton, Derbyshire.*
Bancroft Shed (Bancroft Engine Trust), Gillians Lane, Barnoldswick, Colne, Lancashire. Telephone: Barnoldswick (0282) 814586.*
Boat Museum, Shropshire Union Canal Docks, Ellesmere Port, Cheshire. Telephone: 051-355 1876.*
British Engineerium, Nevill Road, Hove, East Sussex. Telephone: Brighton (0273) 559583.**
Cefn Coed Coal and Steam Centre, Blaenant Colliery, Crynant, West Glamorgan. Telephone: Crynant (063 979) 556.
Chatterley Whitfield Mining Museum, Tunstall, Stoke-on-Trent. Telephone: Stoke-on-Trent (0782) 813377.
Chauntry Mill (D. Gurteen and Sons Ltd), Haverhill, Suffolk.*
Cheddars Lane Sewage Works (Cambridge Museums), Cambridge.*
Coldharbour Mill, Uffculme, near Cullompton, Devon. Telephone: Cullompton (0884) 40960.*
Coleham Pumping Station (Shrewsbury Museums), Shrewsbury, Salop. Telephone: Shrewsbury (0743) 62947.

Combe Sawmill (Combe Mill Society), Combe, near Bladon, Oxford.*
Crofton Pumping Station (The Crofton Society), near Great Bedwyn, Wiltshire. Telephone: Burbage (0672) 810575.*
Dogdyke Pumping Station (Trust), Bridge Farm, Tattershall, Lincolnshire.*
Eastney Pumping Station (Portsmouth Museums), Portsmouth. Telephone: Portsmouth (0705) 827261.*
East Pool Mine (National Trust), Camborne, Cornwall.
Glenruthven Mill, Abbey Road, Auchterarder, Perthshire. Telephone: Auchtermuchty (033 72) 8826.*
Kew Living Steam Museum, Kew Bridge Pumping Station, Green Dragon Lane, Brentford, Middlesex. Telephone: 01-568 4757.**
Leawood Pumping Station (Cromford Canal Society), Cromford, Derbyshire. Telephone Wirksworth (062 982) 3727.*
Locke's Brosna Distillery, Kilbeggan Town Development Trust, Kilbeggan, County Westmeath, Eire.
Middleton Top Winding Engine (Derbyshire Archaeological Society), Wirksworth, Derbyshire. Telephone: Wirksworth (062 982) 3204.
Mill Meece Pumping Station (Trust), near Eccleshall, Staffordshire. Telephone: Stafford (0785) 53734.*
Papplewick Pumping Station (Papplewick Trust), Longdale Lane, Ravenshead, Nottingham. Telephone: Nottingham (0602) 632938.*
Pinchbeck Marsh Pumping Station (Welland and Deeping Drainage Board), Spalding, Lincolnshire.
Ryhope Engines Museum (Trust), Ryhope, Sunderland, Tyne and Wear. Telephone: Sunderland (0783) 210235.*
Scottish Mining Museum: Prestongrange Colliery, Prestonpans, East Lothian. Telephone: 031-665 9904; Lady Victoria Colliery, Newtongrange, Midlothian. Telephone: 031-663 7519.
Springhead Pumping Station Museum (Yorkshire Water Authority, Eastern Division), Hull (by appointment only).
Stamford Brewery Museum, All Saints Street, Stamford, Lincolnshire. Telephone: Stamford (0780) 52186.
Stott Park Bobbin Mill, (English Heritage), Newby Bridge, Cumbria.
Stretham Old Engine, Stretham, near Ely, Cambridgeshire.
Tees Cottage Pumping Station (Trust), Coniscliffe Road, Darlington.*
Tower Bridge Museum, London E1.
Trencherfield Mill, Wigan Pier, Wigan, Lancashire WN3 4EU. Telephone: Wigan (0942) 323666.**
Upper Cherry Gardens Waterworks (Folkestone and District Water Company), Folkestone, Kent.*
Washington F Pit, Washington New Town, Tyne and Wear.
Westonzoyland Pumping Station (Trust), Westonzoyland, near Bridgwater, Somerset. Telephone: Taunton (0823) 412713.*